The Worthy Nuisance

*The Subterranean Arterial Network (SAN)
and the End of the Highway Crisis*

The Worthy Nuisance

The Subterranean Arterial Network (SAN) and the End of the Highway Crisis

Mahendra Jagir

Studio of Books LLC
5900 Balcones Drive Suite 100
Austin, Texas 78731
www.studioofbooks.org
Hotline: (254) 800-1183

Ordering Information:
Special discounts are available on quantity purchases by corporations, associations, and others. For details, contact the publisher at the address above.

Printed in the United States of America.

ISBN-13: Paperback 978-1-970283-59-4
          Hardback   978-1-970283-60-0
          eBook   0978-1-970283-61-7

Library of Congress Control Number: 2026907481

Dedication

To my grandchildren: Dustin, Matthew, Georgia, and Jolie.
This is the blueprint for the world you deserve—one where the heavy
lifting is done in silence, leaving the surface for you to explore.

PREFACE

The Burden of Progress

For much of modern history, we have accepted a quiet trade.

In exchange for progress, we have tolerated disruption.

We have accepted that the movement of goods—so essential to our daily lives—must come with noise, congestion, pollution, and strain. That highways must be crowded, that cities must endure the constant presence of heavy freight, and that infrastructure must continuously fight against the very systems it was designed to support.

We have normalized this condition.

We have learned to live with it.

But normalization is not the same as necessity.

The Illusion of Efficiency

At first glance, the current system appears efficient.

Goods move from ports to cities. Stores remain stocked. Supply chains operate across vast distances. The machinery of global trade functions with remarkable scale.

But efficiency is not measured by movement alone.

It is measured by how that movement affects everything around it.

And when viewed through this lens, the system reveals its cost.

Congestion delays not only freight, but people.

Roads deteriorate under constant pressure.

Communities adapt to noise and pollution.

What we call efficiency is often a system compensating for its own limitations.

A System Built for a Different Time

The infrastructure we rely on today was not designed for the demands it now carries.

Highways were built for regional travel. Cities were planned around human movement. Roads were designed for vehicles of a different scale and purpose.

But over time, these systems were asked to do more.

Global trade expanded. Freight volumes increased. The demands placed on infrastructure grew beyond its original intent.

Yet instead of rethinking the system, we expanded it.

We widened roads.

We added lanes.

We reinforced structures.

We adapted the surface to carry a burden it was never meant to bear.

The Question We Avoided

For decades, the focus has been on managing the problem.

How do we reduce congestion?

How do we improve traffic flow?

How do we repair infrastructure more efficiently?

But rarely have we asked a more fundamental question:

What if the problem is not how we manage the system—but where the system exists?

A Different Perspective

The idea behind the Subterranean Arterial Network did not begin with technology.

It began with a shift in perspective.

Instead of asking how to improve surface logistics, it asked whether surface logistics should exist at all—at least in its current form.

What if the movement of heavy freight could occur somewhere else?

What if the burden could be relocated, rather than managed?

What if the surface could be freed?

The Space Beneath Us

Beneath every road, every city, every landscape, there is space.

Not empty space—but available space.

A dimension of infrastructure that has remained largely unused for logistics, despite its potential to transform how systems operate.

This is where the SAN begins.

Not as an expansion of what exists, but as a relocation of what burdens it.

The Philosophy of Removal

The SAN is not built on adding more.

It is built on removing what no longer belongs.

Removing heavy freight from public roads.

Removing disruption from daily life.

Removing the conflict between infrastructure and the people it serves.

It is not about building larger systems.

It is about building smarter ones.

A Vision for Balance

This book is not only about engineering or logistics.

It is about balance.

A balance between movement and stillness.

Between progress and preservation.

Between what is necessary and what is acceptable.
It is an exploration of how systems can evolve—not by expanding outward, but by integrating more thoughtfully with the world around them.

For the Next Generation

At its core, this work is about the future.
For Dustin, Matthew, Georgia, and Jolie, it represents a different way of thinking about progress.

A world where infrastructure does not dominate the environment.

Where systems serve without overwhelming.

Where growth does not come at the cost of quality of life.

It is a vision of a world where the burdens we have accepted are no longer necessary.

The Beginning of a Shift

The Subterranean Arterial Network is not presented as a final answer.

It is presented as a beginning.

A new way of approaching infrastructure.
A new way of thinking about movement.
A new way of defining progress.

It challenges long-standing assumptions and offers an alternative—one that is quieter, more efficient, and more sustainable.

A Simple Idea

At its simplest, the idea is this:

We did not need to move more.

We needed to move differently.

The Burden We Can Release

The systems we build shape the world we live in.

For too long, we have accepted systems that require compromise—systems that function, but at a cost.

The SAN proposes something different.

A system that functions without burden.

A system that supports without disrupting.

A system that carries the weight of progress—so that we no longer have to.

Table of Contents

Preface: The Burden of Progress 1

Introduction: the Surface Paradox 7

Chapter 1: The SAN Philosophy 13

Chapter 2: Engineering the Void (Resilience & Maglev) 21

Chapter 3: The Golden State Sub-Arterial (GSSA) Model 29

Chapter 4: The SHIELD Integration (The Sprinkler Logic) 37

Chapter 5:
Safety System — SHIELD Integrated Shutdown (SISS) 45

Chapter 6:
Economics of the Underground (The Wealth of Silence) 55

Chapter 7: Revenue, ROI, and Societal Impact 63

Chapter 8: Global Scalability (From Fiji to the World) 71

CONCLUSION: The Silent Renaissance 79

APPENDICES: 85

INTRODUCTION

The Surface Paradox

A System in Plain Sight

There is a paradox that defines the modern world—one so visible that it often goes unnoticed.

We have built a global system capable of moving goods across oceans, continents, and cities with remarkable speed and scale. It is a system that sustains economies, supports industries, and ensures that resources reach the people who need them.

And yet, the final stage of this system—the movement of goods across land—remains one of its greatest weaknesses.

Every day, highways carry the weight of global trade.

Massive freight trucks move alongside passenger vehicles, sharing the same space, the same lanes, and the same infrastructure. What should be a coordinated system instead becomes a point of conflict.

This is the surface paradox:

The very system that enables progress also disrupts it.

The Weight of Movement

At the center of this paradox lies a simple imbalance.

An 80,000-pound freight truck travels on the same road as a 3,000-pound passenger car.

This difference is not just numerical—it is structural.

Each heavy truck exerts a level of stress on infrastructure that far exceeds that of lighter vehicles. Roads designed for general transportation are forced to carry loads they were never intended to support.

The consequences are unavoidable.

Roads deteriorate.

Bridges weaken.

Maintenance becomes constant.

The system does not stabilize.

It strains.

The Cost of Coexistence

The mixing of freight and passenger traffic creates more than physical stress—it creates operational conflict.

Freight vehicles move differently. They accelerate more slowly, require more space, and operate under different constraints. Passenger vehicles, by contrast, are designed for flexibility and responsiveness.

When these two systems coexist on the same infrastructure, inefficiency emerges.

Traffic slows.

Delays increase.

Predictability declines.

The result is a system that functions—but not optimally.

We compensate for this inefficiency through expansion, regulation, and adaptation.

But the underlying conflict remains.

Managing the Problem Instead of Solving It

For decades, the response to congestion and infrastructure strain has followed a familiar pattern.

We widen highways.

We add lanes.

We improve traffic management systems.

These solutions provide temporary relief.

But they do not resolve the fundamental issue.
As capacity increases, usage follows. New lanes fill. Congestion returns. The cycle continues.
This is not failure—it is limitation.
The system is operating as designed.
The problem is that the design itself has not evolved.

A Question of Placement

What if the issue is not how we move goods—but where we move them?

This question shifts the conversation entirely.

Instead of focusing on improving surface logistics, it challenges the assumption that surface logistics is the appropriate solution.

The surface is a shared space.

It is used for transportation, habitation, commerce, and community. It is where human activity takes place.

Introducing heavy industrial movement into this space creates conflict—not because the movement is unnecessary, but because the placement is incompatible.

The Limits of the Surface

Surface infrastructure operates within constraints.

It is exposed to weather.

It competes for space.

It is influenced by human behavior.

These factors introduce variability—conditions that cannot be fully controlled.

Even with advanced systems, surface logistics remains subject to interruption.

Traffic congestion cannot be eliminated entirely.

Weather cannot be controlled.

Human variability cannot be removed.

The system can be improved.

But it cannot be perfected.

The Opportunity Beneath

Beneath the surface lies a different environment.

One that is protected from weather.

One that is free from congestion.

One that operates independently of human variability.

It is a space where movement can occur under controlled conditions.

A space where systems can be designed for efficiency rather than compromise.

This is where the SAN begins.

Relocating the Burden

The concept of the Subterranean Arterial Network is built on a simple but transformative idea:
Relocate the burden.
Instead of forcing heavy freight onto public roads, move it into a dedicated underground system. Separate industrial logistics from human mobility.
This separation resolves the conflict at its source.
Passenger traffic becomes smoother and more predictable.

Freight movement becomes continuous and efficient.

Infrastructure is preserved.

The system is no longer strained.

It is balanced.

A Shift in Thinking

The SAN does not attempt to improve the existing system.

It redefines it.

It challenges long-standing assumptions about how infrastructure should be designed and where it should exist.

It introduces a new perspective:

That efficiency is not achieved by adding more to the surface—but by rethinking how systems are organized.

The Beginning of a New Layer

The SAN represents the creation of a new layer of infrastructure.

A layer that operates beneath the surface, supporting the systems above without interfering with them.

This layer is not visible.

But its impact is.

Congestion decreases.

Noise is reduced.

Air quality improves.

The surface becomes a space for people again.

From Conflict to Coordination

When freight and passenger systems are separated, they no longer compete.

They coordinate.

Each operates within its own environment, optimized for its specific function.

This coordination creates efficiency—not through control, but through design.

The Path Forward

The purpose of this book is not only to present a system.

It is to present a shift.

A shift from managing problems to resolving them.

A shift from expanding limitations to redefining possibilities.

The Subterranean Arterial Network is one expression of this shift.

It is a system that asks us to reconsider what we accept as necessary—and what we can change.

What Comes Next

The chapters that follow explore this system in detail.

They examine its philosophy, its engineering, its economic impact, and its global potential.

They present not only how the SAN works—but why it matters.

Because the challenge we face is not simply one of logistics.

It is one of design.

The Surface, Reimagined

The surface does not need to carry everything.

It was never meant to.

By rethinking how we use it—and what we place upon it—we can restore balance to the systems that support our lives.

A Simple Realization

In the end, the surface paradox can be resolved with a simple realization:

We did not need to move less.

We needed to move differently.

CHAPTER 1

The SAN Philosophy

Defining the "Worthy Nuisance"

For over a century, humanity has lived with a contradiction so familiar that it has become nearly invisible.

The movement of goods—food, medicine, materials, and technology—has been the foundation of modern civilization. It is the unseen engine that powers economies, sustains populations, and connects distant regions into a single, interdependent system. Every item we use, every service we depend on, and every market we participate in is supported by the continuous and coordinated movement of freight.

Yet, while this system enables modern life, it also disrupts it.

We see it in the endless lines of freight trucks stretching across highways, often moving slowly, sometimes barely moving at all. We hear it in the constant hum and rumble of engines cutting through cities and neighborhoods. We feel it in the wear of our roads, the delays in our commutes, and the growing tension between infrastructure and the people it is meant to serve.

This is the "nuisance."

And yet, it is a worthy one.

It exists because global trade exists. It persists because society depends on it. It represents the success of a system that has enabled unprecedented access, growth, and connectivity. But the way in which we have chosen to manage this success reveals a deeper problem.

The nuisance is not the movement of goods.

The nuisance is the system we use to move them.

The Conflict of Two Eras

We are attempting to move 21st-century demand on a 20th-century foundation.

Modern logistics operates at a scale that would have been unimaginable just decades ago. Containerized shipping has revolutionized global trade, allowing goods to move efficiently across oceans and continents. Supply chains have become faster, more complex, and more interconnected than ever before.

But when these goods arrive on land, they are transferred into a system that has not evolved at the same pace.

Highways—originally designed for regional travel and light passenger vehicles—have become the primary channels for global freight distribution. On these roads, an 80,000-pound freight truck shares space with a 3,000-pound family car.

This imbalance is not simply inefficient—it is fundamentally flawed.

Each heavy truck places immense stress on infrastructure, accelerating wear and reducing the lifespan of roads and bridges. The cost of maintenance rises, repairs become more frequent, and disruptions become routine.

At the same time, the coexistence of freight and passenger traffic creates congestion, unpredictability, and increased risk. The system becomes slower, less reliable, and more expensive to sustain.

This is not a failure of effort or investment.

It is a failure of alignment.

A Problem of Geometry, Not Effort

For decades, the response to this challenge has been consistent: expand.

When roads become congested, we widen them. When traffic increases, we add lanes. When systems begin to fail, we repair them and extend them further.

But this approach is inherently limited.

Surface space is finite. Every expansion consumes land, increases cost, and introduces new layers of complexity. And perhaps most critically, expansion often leads to increased usage—filling new capacity as quickly as it is created.

This cycle of horizontal growth is reactive, not transformative.

It addresses symptoms without resolving the underlying issue.

The SAN philosophy begins with a different question:

What if the problem is not the amount of space—but how we use it?

What if the solution is not to build outward, but to build differently?

From Horizontal Expansion to Vertical Integration

Beneath the surface lies an opportunity that has largely gone unexplored.

The Earth itself provides space—vast, stable, and capable of supporting infrastructure that operates independently of surface limitations. It is an environment where movement can occur without congestion, without weather interference, and without competing with human activity.

The SAN introduces the concept of **vertical integration** in infrastructure.

Instead of concentrating all movement on a single plane, it separates systems based on their purpose.

The surface becomes a space for people—mobility, community, and environment.

The underground becomes a space for industry—freight, energy, and logistics.

This separation is not merely an improvement.

It is a redefinition of how systems coexist.

By removing conflict from the surface, both layers are allowed to function more effectively. The result is not only increased efficiency, but restored balance.

The Three Pillars of the SAN

The Subterranean Arterial Network is built upon three foundational principles, each addressing a core limitation of modern infrastructure.

1. Segregation of Mass

The mixing of vastly different weight classes is one of the most significant inefficiencies in transportation today.

Heavy freight vehicles share the same roads as light passenger vehicles, creating structural strain, safety risks, and operational conflict.

The SAN resolves this by separating these systems entirely.

Freight is relocated to a dedicated underground environment, where it can operate without interference. Passenger vehicles reclaim the surface, free from the disruption of heavy logistics.

This is not a redistribution of traffic.

It is the elimination of conflict.

2. Velocity of the Pulse

Surface transportation is inherently inconsistent.

Traffic signals, congestion, accidents, weather conditions, and human variability create a system defined by interruption. Movement is fragmented, unpredictable, and often inefficient.

The SAN replaces this with a system defined by continuity.

Cargo moves through a controlled environment where external disruptions do not exist. It operates like a pulse—steady, rhythmic, and uninterrupted.

This consistency transforms logistics from a reactive system into a reliable one.

Speed is no longer measured by peak performance, but by predictability.

3. Infrastructure Preservation

Modern infrastructure is not failing because it is poorly designed.

It is failing because it is overburdened.

Heavy freight traffic accelerates the degradation of roads and bridges, forcing continuous maintenance and repair. This cycle is costly, disruptive, and unsustainable.

By removing the primary source of stress—the crushing load of freight—the SAN allows infrastructure to return to its intended function.

Roads last longer.

Maintenance becomes less frequent.

Public investment becomes more effective.

The system is no longer under constant pressure.

It is allowed to endure.

The Silent Renaissance

The ultimate goal of the SAN is not visibility.

It is invisibility.

We envision a world where logistics are no longer a dominant presence in daily life. A world where goods move efficiently without disrupting the spaces people inhabit.

In this world:

Highways are quieter.

Cities are less congested.

Air is cleaner.

Communities are more livable.

The machinery of global trade continues to operate—but it does so beneath awareness.

This is the **Silent Renaissance.**

A transformation not defined by what is added, but by what is removed.

A Shift in Perspective

The SAN challenges a deeply rooted assumption:

That progress must be visible.

For generations, infrastructure has been associated with expansion—larger roads, more vehicles, greater presence. But visibility is not always a measure of success.

Sometimes, the most effective systems are those that disappear.

Systems that perform their function so seamlessly that they become part of the environment, rather than a disruption to it.

The SAN represents this shift.

It is infrastructure that supports without intruding.

That functions without dominating.

That exists without overwhelming.

The Human Dimension

Beyond efficiency and engineering, the SAN addresses something often overlooked in infrastructure design:

The human experience.

Surface logistics does not only affect transportation—it affects how people live.

Noise impacts communities.

Congestion affects daily routines.

Pollution influences health.

By removing heavy freight from the surface, the SAN improves not only systems, but lives.

Cities become more navigable.

Neighborhoods become quieter.

Public spaces become more accessible.

Infrastructure returns to serving people, rather than competing with them.

A Legacy Beneath the Surface

This philosophy is not only about solving present challenges.

It is about shaping the future.

For Dustin, Matthew, Georgia, and Jolie, this vision represents a world where progress is no longer defined by visible burden.

A world where systems are designed with foresight, responsibility, and balance.

A world where infrastructure supports life without diminishing it.

It is a legacy not of expansion, but of refinement.

The Beginning of a New Framework

The SAN is more than a proposal.

It is a new framework for thinking about infrastructure.

It asks us to reconsider long-held assumptions:

That surface space is the only viable space.

That growth must be visible.

That efficiency requires compromise.

It offers an alternative:

A system where movement is continuous, but unobtrusive.

Where infrastructure is powerful, but quiet.

Where progress is achieved not by adding more—but by designing better.

A World Reimagined

The "Worthy Nuisance" is not eliminated.

It is transformed.

The movement of goods continues—essential, constant, and necessary. But it no longer defines the spaces we inhabit.

It no longer disrupts the surface.

Instead, it operates beneath it—quietly, precisely, and invisibly.

And in that transformation lies the true significance of the SAN:

Not simply as an innovation, but as a redefinition of how progress itself is achieved.

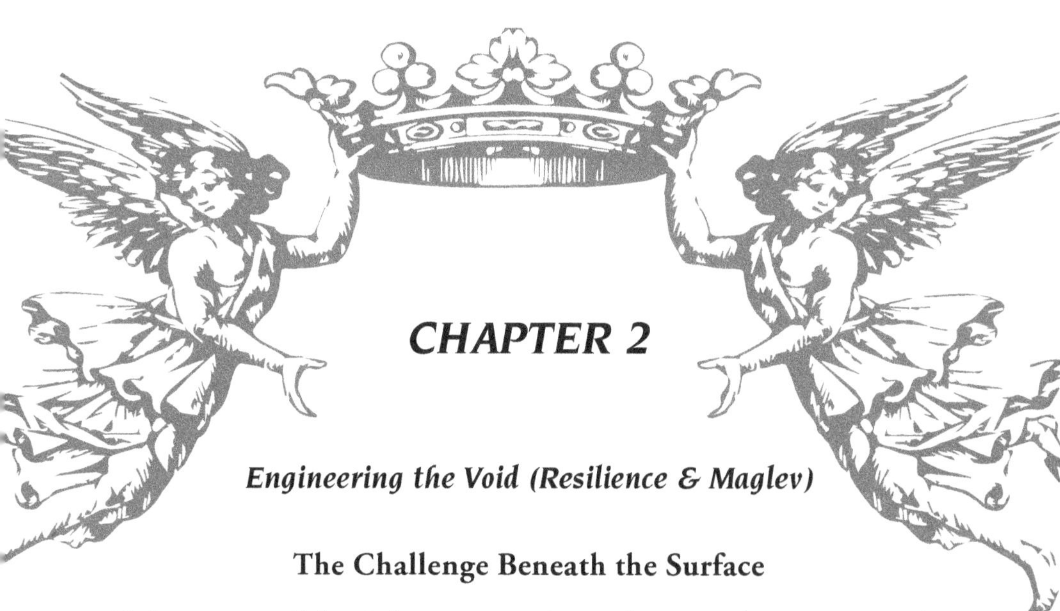

CHAPTER 2

Engineering the Void (Resilience & Maglev)

The Challenge Beneath the Surface

If the vision of the Subterranean Arterial Network is to move the weight of the world beneath our feet, then the question becomes immediate and unavoidable:

Can the Earth safely carry that burden?

The surface, for all its complexity, is familiar territory. We build on it, expand across it, and repair it when it fails. Its challenges—weather, traffic, wear—are visible and understood.

The underground, however, presents a fundamentally different environment.

It is a domain shaped by pressure, instability, and constant motion. The Earth is not static; it shifts, settles, and responds to forces both gradual and sudden. Beneath the surface, these forces are not obstacles to be avoided—they are conditions to be embraced.

To build within this environment is not simply an engineering exercise.

It is a negotiation with the planet itself.

The SAN does not attempt to dominate the Earth.

It is designed to coexist with it.

Rethinking Strength: Flexibility Over Rigidity

Traditional infrastructure often relies on a simple principle: strength through rigidity.

Bridges are reinforced. Roads are layered. Tunnels are constructed as continuous, solid structures intended to resist external forces.

But underground, rigidity becomes a liability.

When the Earth moves—as it inevitably will—a rigid structure cannot adapt. It fractures, cracks, and ultimately fails.

A rigid tunnel is a brittle tunnel.

The SAN approaches this challenge differently.

Instead of resisting movement, it accommodates it.

Rather than building a single, continuous structure, the SAN employs a **segmented architecture**—a system composed of individual sections that function together as a flexible whole.

Each segment is designed to connect with the next through specialized joints—what can be described as **"accordion joints."**

These joints incorporate high-elasticity materials capable of controlled expansion and contraction. When the surrounding ground shifts, the tunnel does not resist the movement—it absorbs it.

The structure bends, adjusts, and stabilizes.

It moves with the Earth, rather than against it.

Seismic Resilience in Motion

Nowhere is this approach more critical than in regions prone to seismic activity.

In areas such as California, earthquakes are not hypothetical events—they are certainties. Any underground system must be capable of enduring repeated stress without catastrophic failure.

The SAN addresses this through layered resilience.

First, the segmented design allows localized movement without compromising the entire structure. A shift in one section does not propagate into a system-wide failure.

Second, **viscous dampers** are integrated between the tunnel walls and the internal rail systems. These function as shock absorbers, isolating the core transport components from external vibrations.

During a seismic event, the tunnel structure may experience movement—but the internal system remains stable.

Cargo flow can slow, pause, or reroute safely, but it does not derail.

This distinction is critical.

The goal is not to prevent movement.

The goal is to ensure stability within movement.

Stability in Unstable Ground

Beyond seismic activity, the underground presents another significant challenge: **soil instability**, particularly in regions susceptible to liquefaction.

In these conditions, saturated soil can lose its structural integrity during seismic events, behaving more like a liquid than a solid. Structures built without proper stabilization can shift, tilt, or even float.

The SAN addresses this through proactive engineering.

During construction, **deep-grout injection** is used to reinforce the surrounding soil. As tunnel boring progresses, stabilizing agents are introduced into the ground, creating a solidified envelope around the structure.

This reinforced zone acts as a protective barrier, ensuring that the tunnel remains anchored even under extreme conditions.

In addition, the system incorporates **buoyancy control mechanisms**. By integrating utility conduits and structural weighting into the design, the tunnel maintains equilibrium, preventing upward displacement during liquefaction events.

The result is a structure that is not only stable—but secure.

Eliminating Friction: The Maglev Advantage

Once structural resilience is established, the next challenge is efficiency.

Traditional freight systems rely on mechanical contact—wheels against rails, engines against resistance. This contact generates friction, wear, and energy loss.

Over time, it leads to degradation.

The SAN eliminates this limitation entirely.

At the heart of its transport system is **magnetic levitation**, or maglev technology.

Instead of rolling along tracks, cargo platforms—known as sleds—are lifted and propelled using electromagnetic forces. There is no physical contact between the moving components and the track.

This creates a fundamentally different system.

- No friction between surfaces

- Minimal mechanical wear

- Reduced maintenance requirements

- Smoother, more stable motion

The propulsion system—based on **Linear Induction Motors (LIM)**—activates only in the segments where movement is required. As a sled passes through a section, that segment powers its motion, then deactivates once the sled moves on.

This localized activation conserves energy and enhances efficiency.

The system does not push continuously.

It responds precisely.

Redefining Speed

In conventional logistics, speed is often constrained by external variables.

Traffic, weather, and infrastructure limitations all influence how quickly goods can move. Even the most advanced surface systems are subject to interruption.

The SAN operates in a controlled environment where these variables are eliminated.

This allows for sustained high-speed movement—without compromise.

Cargo moves at consistent velocities, unaffected by congestion or environmental conditions. Travel times become predictable. Delivery schedules become reliable.

Speed is no longer a peak condition.

It becomes a constant.

The Thin Air Strategy

Even with friction eliminated, another force remains: air resistance.

At high speeds, air becomes a significant barrier. It creates drag, increases energy consumption, and limits efficiency.

To address this, the SAN introduces a controlled **low-pressure environment** within its tunnels.

Rather than attempting a full vacuum—which presents significant technical challenges over long distances—the system reduces air density to a fraction of normal atmospheric levels.

This is achieved through a network of large-scale pumping systems, integrated with the SAN's energy infrastructure.

The impact is profound.

By reducing air resistance, the system significantly lowers the energy required to move heavy cargo. Containers weighing tens of tons can travel at high speeds with efficiency comparable to much lighter vehicles on the surface.

This is not merely an improvement.

It is a transformation of how energy is used in transportation.

The Intelligence of the System

A system of this complexity cannot rely solely on physical design.

It must also be intelligent.

Embedded throughout the SAN is a network of sensors and monitoring systems that continuously analyze conditions within the environment.

These systems detect:

- Structural changes

- Temperature variations

- Pressure fluctuations

- Mechanical irregularities

But detection alone is not enough.

The system must respond.

The SHIELD Integrated Shutdown System (SISS)

The SAN's response mechanism is governed by the **SHIELD Integrated Shutdown System (SISS)**—a distributed safety network designed to isolate and resolve issues without disrupting the entire system.

The principle is simple:

Respond locally.

Maintain globally.

When a potential issue is detected, the affected segment is isolated. Power is temporarily reduced or redirected. Containment systems activate, ensuring that the issue remains confined.

At the same time, cargo flow is rerouted through parallel pathways, allowing the system to continue operating.

This approach ensures continuity.

The system does not fail as a whole.

It adapts in parts.

Predictive Maintenance: Preventing Failure Before It Begins

One of the most significant advantages of the SAN is its ability to anticipate problems before they occur.

Through continuous data analysis, the system identifies patterns—subtle changes that indicate potential future issues.

A slight increase in vibration.

A gradual rise in temperature.

A minor deviation in alignment.

These signals trigger maintenance actions before failure occurs.

Autonomous repair units are deployed during low-traffic periods, addressing issues without interrupting operation.

This transforms maintenance from a reactive process into a proactive one.

The system does not wait for failure.

It prevents it.

Designed for Generations

The true measure of infrastructure is not how it performs in the present, but how it endures over time.

The SAN is designed with a lifespan measured in generations.

Protected from weather, shielded from surface stress, and engineered for adaptability, it operates within an environment that preserves its integrity.

Unlike roads that crack under pressure or bridges that corrode with exposure, the SAN exists in a controlled space—one where degradation is minimized and longevity is maximized.

This is infrastructure that is not rebuilt repeatedly.

It is maintained continuously.

The Invisible Precision

The greatest achievement of the SAN's engineering is not its complexity.

It is its subtlety.

The system performs at a high level, yet remains largely unseen. It operates with precision, yet does not intrude upon daily life.

Its success is measured not by its visibility, but by its absence.

The absence of congestion.

The absence of disruption.

The absence of failure.

A Foundation Beneath the Future

For Dustin, Matthew, Georgia, and Jolie, this chapter represents something more than technical innovation.

It represents certainty.

The certainty that the systems supporting their world are designed to last. That infrastructure can be both powerful and unobtrusive. That progress can be achieved without compromise.

Beneath their feet lies a system engineered not only for performance, but for permanence.

A system that carries the weight of the world—quietly, reliably, and without interruption.

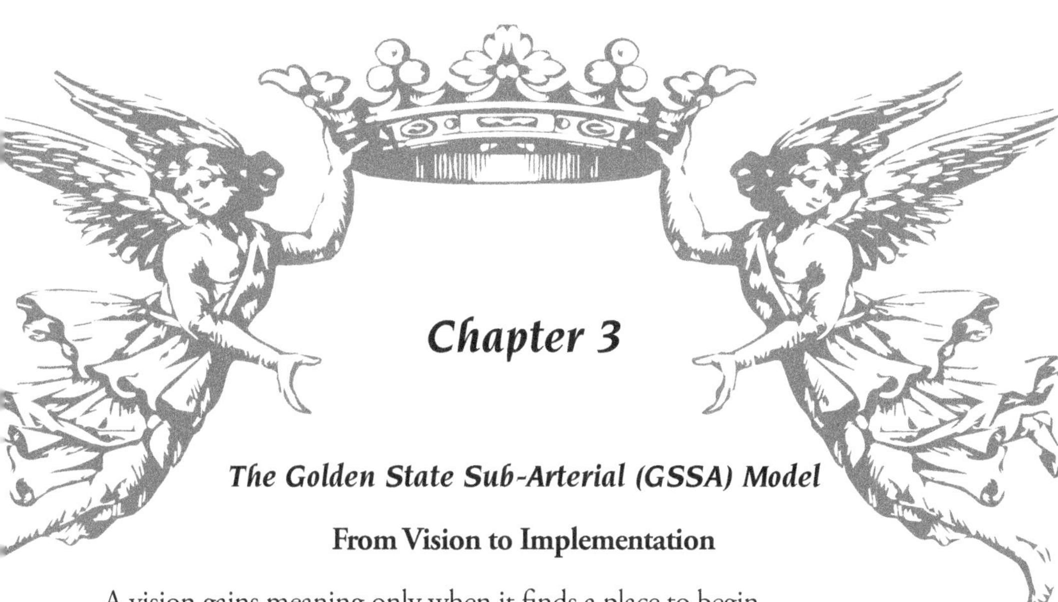

Chapter 3

The Golden State Sub-Arterial (GSSA) Model

From Vision to Implementation

A vision gains meaning only when it finds a place to begin.

The Subterranean Arterial Network is not intended to remain theoretical. It is designed to be built, tested, and proven within a real-world environment—one that reflects both the challenges and opportunities of modern infrastructure.

That environment is California.

Few regions illustrate the pressures of global logistics as clearly. Its ports serve as gateways for international trade. Its highways carry immense volumes of freight. Its cities experience the daily effects of congestion, pollution, and infrastructure strain.
Here, the "Worthy Nuisance" is not abstract.

It is visible in the lines of trucks waiting at ports.

It is measurable in the cost of road maintenance.

It is experienced in the time lost to congestion.

The Golden State Sub-Arterial (GSSA) represents the first full-scale application of the SAN philosophy—a system designed to transform one of the most complex logistics corridors in the world.

It is not simply a pilot.

It is a blueprint.

A System Modeled on Life

To understand the GSSA, it is useful to move beyond traditional engineering analogies and consider a different framework—one rooted in biology.

The system is designed as a living network.

The tunnels function as arteries.

The hubs act as organs.

The movement of cargo becomes the pulse.

This is not a metaphor for simplicity.

It is a model for efficiency.

In biological systems, each component performs a specific function while contributing to the stability of the whole. The same principle applies here. Every part of the GSSA is designed to operate independently while remaining fully integrated.

The result is a system that is both resilient and adaptable—capable of sustaining continuous flow without interruption.

The Port Interface: The Beginning of Flow

The journey of goods begins at the ports.

Traditionally, this is where the first disruption occurs. Containers are unloaded from ships and transferred onto trucks, immediately introducing them into congested urban environments.

The GSSA replaces this step with a direct transition into the subterranean system.

At the center of this transition are **Descending Freight Wells**—vertical shafts that connect port operations directly to the underground network.

These wells function as the "mouth" of the system.

Instead of lifting containers onto surface vehicles, cranes lower them directly onto maglev sled platforms positioned within these shafts. From there, the cargo enters the SAN without ever interacting with surface traffic.

Before entering the main network, containers are sorted using advanced logistics systems. Each unit is assigned a destination hub, ensuring that once it begins its journey, it moves without interruption.

There are no intersections.

No delays.

No uncertainty.

Only flow.

The Arterial Network: Movement Without Interruption

Once within the system, cargo enters the arterial network—the primary channels through which goods move across the region.

These tunnels run beneath existing transportation corridors, aligning with established routes while removing the physical burden from the surface.

Within these tunnels, maglev sleds transport cargo at high speed, maintaining a steady and uninterrupted flow.

Unlike surface systems, where movement is influenced by countless external factors, the SAN operates within a controlled environment.

There is no congestion.

No weather interference.

No human variability.

Movement becomes consistent.

This consistency redefines logistics.

Delivery times become predictable.

Supply chains become stable.

Efficiency becomes inherent.

The system does not accelerate and decelerate.

It flows.

Regional Hubs: Centers of Distribution

If the tunnels enable movement, the hubs enable distribution.

Rather than relying on a single port to serve an entire region, the GSSA distributes freight through strategically placed inland hubs. These hubs are located at distances that balance efficiency and accessibility—typically between 100 and 300 miles from the coast.

This decentralization reduces pressure on coastal infrastructure and creates a more balanced system.

Three primary hubs define the GSSA model:

The Victorville Gateway (The Southern Valve)

The Victorville hub serves as the first major distribution point for freight entering from Southern California ports.

Located in an area with ample space and lower population density, it provides an ideal environment for large-scale sorting and redistribution.

Here, containers are lifted from the underground system using electromagnetic elevators and transferred into staging areas.

The advantage of this location lies in its capacity.

Unlike congested port cities, Victorville allows for efficient organization and redistribution of goods. Containers destined for nearby regions are processed locally, while others are directed deeper into the network.

This hub acts as a pressure release point—reducing congestion at the coast and redistributing it inland in a controlled manner.

The Fresno Agricultural Hub (The Central Spine)

Fresno represents a different kind of hub—one defined not only by distribution, but by preservation.

As a central point within California's agricultural region, this hub integrates **cold-chain logistics** into the SAN.

Perishable goods are introduced into the system through climate-controlled modules, maintaining consistent conditions throughout transport.

On the surface, agricultural products are often exposed to extreme temperatures, leading to spoilage and loss. Within the SAN, these goods move through a stable environment, preserving quality and extending shelf life.

This is not simply an efficiency gain.

It is a transformation of how resources are protected and delivered.

The Stockton/Delta Hub (The Northern Valve)

The Stockton hub serves as a major distribution center for Northern California and the Bay Area.

Here, the primary flow of cargo is divided into smaller channels—what can be described as capillaries.

These branches extend toward key regions, allowing goods to reach their destinations with precision and efficiency.

This layered distribution system ensures that volume is managed effectively, preventing congestion within the network itself.

The system does not rely on a single pathway.

It distributes intelligently.

The Final Mile: Completing the Journey

No logistics system is complete without addressing the final step—delivery to the end destination.

The GSSA resolves this through the **50-Mile Rule**.

No container traveling through the SAN is transported more than 50 miles on the surface.

By the time goods emerge from the underground network, they are already close to their destination. From there, they are transferred onto lightweight, electric vehicles designed for short-haul delivery.

This approach eliminates long-distance trucking, significantly reducing infrastructure strain and environmental impact.

Surface roads are no longer dominated by heavy freight.

They support localized, efficient delivery systems.

Relieving the Surface

The impact of the GSSA is most visible not within the tunnels, but above them.

As freight is removed from highways, the surface begins to change.

Traffic becomes lighter.

Noise levels decrease.

Air quality improves.

Infrastructure that once struggled under constant stress begins to recover.

Communities that were defined by congestion begin to transform.

The surface is no longer a space of compromise.

It becomes a space of possibility.

Scalability Through Design

The GSSA is not a fixed system.

It is designed to grow.

Additional hubs can be introduced. New arterial lines can be extended. Capacity can increase without disrupting existing operations.

This scalability ensures that the system remains relevant as demand evolves.

It is not limited by its initial design.

It adapts.

A Model for the Future

The significance of the GSSA extends beyond its immediate impact.

It demonstrates that the SAN is not only possible, but practical.

It provides a model that can be replicated in other regions—adapted to different geographies, economies, and infrastructure needs.

It shows that a new approach to logistics is not only desirable, but achievable.

A Legacy of Balance

For Dustin, Matthew, Georgia, and Jolie, the GSSA represents more than a system.

It represents balance.

A balance between movement and stillness.

Between progress and preservation.

Between infrastructure and the environment.

It is a world where the systems that sustain life no longer dominate it.

Where the burdens of industry are carried quietly, beneath the surface.

The Quiet Transformation

The GSSA does not announce itself.

It does not rise above the skyline or reshape the landscape in visible ways.

Its transformation is subtle.

It happens beneath the ground, within systems that operate continuously and efficiently.

And yet, its impact is profound.

The world above becomes quieter.

Cleaner

More livable.

This is the beginning of a new kind of infrastructure.

One that supports without overwhelming.

One that transforms without disrupting.

One that carries the weight of progress—so that the surface no longer has to.

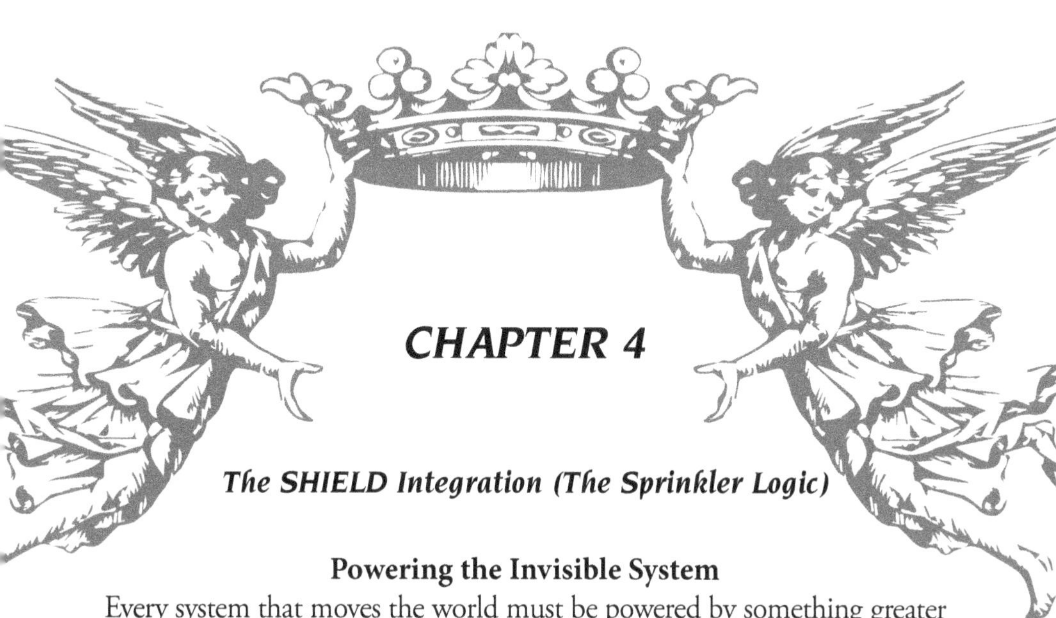

CHAPTER 4

The SHIELD Integration (The Sprinkler Logic)

Powering the Invisible System

Every system that moves the world must be powered by something greater than motion.

It must be sustained.

The Subterranean Arterial Network is not simply a transportation system—it is a continuous operation. Its effectiveness depends not only on how efficiently it moves cargo, but on how reliably it maintains that movement over time.

This introduces a critical question:

What powers a system that never stops?

Traditional infrastructure depends heavily on centralized energy grids—systems that are often complex, strained, and vulnerable to disruption. Power outages, fluctuations, and inefficiencies can ripple through entire networks, halting operations and exposing weaknesses.

For a system like the SAN, such vulnerability is unacceptable.

It cannot stop.

It cannot wait.

It must endure.

The answer lies in integration—not dependence.

The SHIELD Framework

At the core of the SAN's energy strategy is the **SHIELD Project**—a distributed, renewable energy system designed to work in direct coordination with the subterranean network.

The SHIELD framework transforms energy from a supporting element into an integrated component of infrastructure.

It is not an external supply.

It is part of the system itself.

This integration creates a relationship between the surface and the underground that is both functional and symbiotic.

Above, energy is generated.

Below, it is utilized.

Together, they form a continuous loop.

The Surface-to-Subterranean Power Loop

Along major transportation corridors, the surface above the SAN is utilized for renewable energy generation.

Windmills, solar arrays, and other sustainable systems are positioned strategically, forming a network of energy-producing nodes that mirror the structure of the subterranean system below.

These installations are not isolated power sources.

They are connected—feeding energy directly into the SAN through dedicated conduits.

This creates what can be described as an **energy loop**.

Energy is generated at the surface, transmitted below, and used to power the movement of goods. Excess energy can then be redistributed, supporting the broader grid.

Unlike traditional systems, which rely on multiple stages of conversion, the SAN operates primarily on **direct current (DC)**.

This distinction is critical.

Conventional alternating current (AC) systems require repeated conversion between AC and DC, resulting in energy loss. By maintaining a DC-based system, the SAN minimizes these losses, ensuring that more of the generated energy is used effectively.

The result is a system that is not only efficient, but direct.

Energy flows as seamlessly as cargo.

Redundancy as a Principle

In most infrastructure systems, redundancy is treated as a safeguard—something to be activated in the event of failure.

In the SAN, redundancy is fundamental.

The SHIELD network is designed as a series of distributed energy nodes, each capable of functioning independently. These nodes are positioned at regular intervals along the system, creating a decentralized network of power.

If one node experiences disruption, others compensate.

This creates a system that is resilient by design.

Power is not concentrated in a single source.

It is distributed across many.

This distribution ensures that even in the face of localized failure, the system continues to operate.

It does not collapse under stress.

It adapts.

Energy Independence and Stability

One of the defining characteristics of the SHIELD integration is its ability to operate independently of traditional power grids.

While it can connect to external systems, it does not rely on them.

This independence provides stability.

In the event of a regional outage, the SAN continues to function. Cargo continues to move. The system remains operational.

This capability—often referred to as **"limp-home functionality"**—ensures that even under reduced power conditions, active transport units can complete their journeys safely.

Movement is never abruptly halted.

It is managed.

Energy as Infrastructure

In the SAN model, energy is not separate from infrastructure.

It is embedded within it.

The same tunnels that carry cargo also carry power. The same pathways that enable movement also distribute energy. This integration creates efficiencies that extend beyond transportation.

It transforms the SAN into a dual-purpose system:

- A logistics network

- An energy distribution network

This dual functionality enhances the value of the system, allowing it to contribute to broader energy stability while fulfilling its primary role.

Infrastructure becomes more than a means of movement.

It becomes a source of support.

The Sprinkler Logic: A New Approach to Safety

If energy sustains the system, safety ensures its continuity.

Traditional infrastructure often relies on centralized responses to failure. When an issue occurs, entire systems may be slowed, halted, or disrupted.

The SAN adopts a different approach.

Its safety model is inspired by one of the most effective and widely used systems in the built environment: the fire sprinkler.

A sprinkler system does not attempt to address every potential issue simultaneously.

It responds precisely—activating only where needed.

The SAN applies this principle through the **SHIELD Integrated Shutdown System (SISS)**.

The Multi-Sensory Monitoring Network

Embedded throughout the SAN is a network of sensors that continuously monitor conditions within the system.

These sensors operate across multiple dimensions:

• **Thermal monitoring** detects abnormal heat levels

• **Acoustic sensing** identifies structural irregularities

• **Vibration analysis** detects seismic or mechanical disturbances

• **Pressure monitoring** ensures environmental stability

Each segment of the system functions as an independent monitoring unit—a **"digital sprinkler head."**

These units operate continuously, collecting and analyzing data in real time.

They do not wait for failure.

They detect change.

Localized Response, Global Continuity

When an anomaly is detected, the system responds immediately—but with precision.

Instead of shutting down the entire network, only the affected segment is isolated.

Magnetic propulsion within that section is temporarily reduced or deactivated. Containment systems are deployed, sealing the segment and preventing the issue from spreading.

At the same time, cargo flow is redirected through parallel pathways.

Movement continues.

This approach ensures that:

- The issue is contained

- The system remains operational

- Disruption is minimized

The system does not overreact.

It responds intelligently.

Autonomous Intervention

Once a segment is isolated, the next phase begins: resolution.

Autonomous repair units—designed to operate within the subterranean environment—are deployed from the nearest hub.

These units assess the situation, perform necessary repairs, and verify system integrity.

Once conditions are restored, the segment is reintegrated into the network.

This process is seamless.

To the broader system—and to the world above—it is invisible.

The pulse continues uninterrupted.

Predictive Intelligence

The SHIELD integration is not limited to reaction.

It is predictive.

Through continuous data analysis, the system identifies patterns that indicate potential future issues. Minor deviations—imperceptible in isolation—become meaningful when observed over time.

A slight increase in temperature.

A gradual shift in vibration patterns.

A small fluctuation in pressure.

These signals trigger preventive maintenance before failure occurs.

The system learns.

It anticipates.

It evolves.

This is the transition from infrastructure to intelligence.

Guardians Beneath the Surface

For Dustin, Matthew, Georgia, and Jolie, this system represents more than engineering.

It represents protection.

Protection from failure.

Protection from disruption.

Protection from the visible burdens of infrastructure.

For some, this protection is seen in the precision of the system—the seamless flow of energy and movement.

For others, it is felt in what is no longer present—the noise, the congestion, the strain.

The system does not announce itself.

It protects quietly.

The Invisible Guardian

The greatest achievement of the SHIELD integration is not its complexity.

It is its subtlety.

It operates continuously, maintaining stability without drawing attention. It ensures safety without interruption. It supports the system without becoming its focus.

It is infrastructure that protects without intruding.

A System That Sustains Itself

In the end, the SHIELD integration fulfills a critical role:

It ensures that the SAN is not only efficient—but sustainable.

Not only functional—but resilient.

Not only powerful—but reliable.

It creates a system that can operate continuously, adapt to challenges, and maintain its integrity over time.

The Quiet Assurance

The SAN is designed to move the world.

The SHIELD system ensures that it can do so without interruption.

Together, they form a unified system—one that is powered, protected, and sustained from within.

And in that unity lies their strength.

A system that does not depend on external stability—but creates its own.

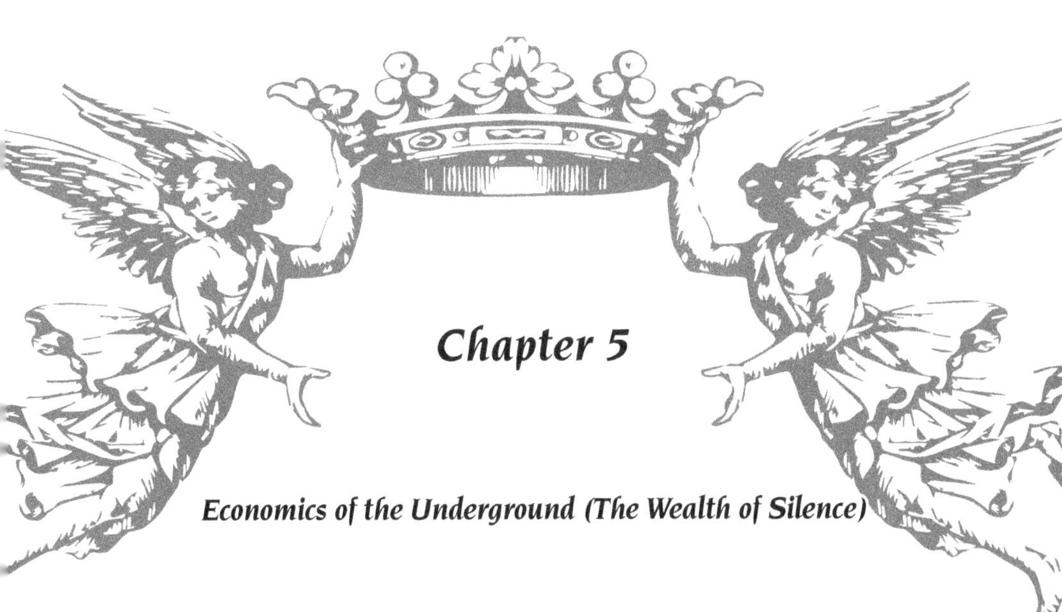

Chapter 5

Economics of the Underground (The Wealth of Silence)

The Cost We No Longer Question

Every system carries a cost.

Some costs are visible—measured in budgets, materials, and labor. Others are less obvious, embedded within inefficiencies, delays, and long-term consequences that accumulate over time.

For generations, the global freight system has been accepted as a necessary expense of progress. Highways are built, repaired, and expanded. Trucks move continuously across regions. Congestion is managed rather than resolved.

These conditions have become normalized.

But normalization does not mean efficiency.

It does not mean sustainability.

And it certainly does not mean that the system we rely on is the best one available.

The truth is simple:

The cost of our current system is far greater than what we see.

The Hidden Burden of Surface Logistics

Surface-based freight transportation is often evaluated in terms of direct expense—fuel, labor, and infrastructure maintenance. But these figures represent only a portion of the total cost.

Beneath them lies a network of indirect impacts.

Heavy trucks accelerate the deterioration of roads and bridges, forcing constant repair and reconstruction. These repairs are expensive, disruptive, and recurring. Infrastructure that should last decades is often rebuilt within years.

Congestion adds another layer of cost.

Time is lost. Fuel is wasted. Deliveries are delayed. Supply chains become unpredictable. Businesses compensate with higher margins, increased inventory, or additional resources—each adding to the overall expense.

Then there is the environmental impact.

Diesel emissions contribute to air pollution, affecting public health and increasing healthcare costs. Noise pollution alters communities. Urban space becomes dominated by the movement of goods rather than the needs of people.

These costs are not isolated.

They are cumulative.

And they are ongoing.

Reframing the Investment

When viewed in isolation, the Subterranean Arterial Network appears to require a significant initial investment.

The scale is large. The engineering is advanced. The construction is complex.

But this perspective is incomplete.

The SAN is not an addition to the current system.

It is a replacement for its most costly component.

The question is not:

"How much does the SAN cost?"

The question is:

"How much does the current system already cost—and for how long?"

When viewed over time, the comparison shifts.

The SAN does not introduce new expense.

It reallocates existing expense into a system that is more efficient, more durable, and more sustainable.

The Structure of Investment

The economic framework of the SAN can be understood through three primary areas of investment.

1. Subterranean Construction

The creation of the underground network represents the largest portion of the investment.

This includes tunnel excavation, structural reinforcement, and the implementation of systems designed to ensure long-term resilience.

Advanced tunnel boring technologies allow for efficient construction across large distances, while reinforced materials ensure durability over decades.

Unlike surface infrastructure, which requires frequent maintenance, the SAN is designed for longevity.

It is not rebuilt repeatedly.

It is built to endure.

2. Transport and Propulsion Systems

The second area of investment focuses on the systems that enable movement within the network.

Maglev propulsion eliminates friction, reducing wear and maintenance costs. Low-pressure environments increase efficiency, lowering energy consumption.

These systems are not only faster—they are more economical over time.

Where traditional systems degrade with use, the SAN maintains performance with minimal intervention.

3. Hubs and Interface Systems

The final component includes the development of ports, descending freight wells, and regional hubs.

These systems serve as the connection points between global trade and the subterranean network.

They are designed to manage flow efficiently—sorting, distributing, and directing cargo without delay.

Together, these three components form a complete system—one that replaces the inefficiencies of surface logistics with a unified and optimized network.

The Economics of Removal

The most significant economic impact of the SAN comes not from what it adds—but from what it removes.

It removes the primary cause of infrastructure degradation.

It removes long-distance freight from highways.

It removes the conditions that create congestion and delay.

This distinction is critical.

The SAN does not optimize the existing system.

It eliminates its greatest inefficiencies.

The Value of Efficiency

Efficiency in logistics is not simply about speed.

It is about reliability.

When goods move without interruption, supply chains become predictable. Businesses can operate with greater precision. Resources are used more effectively.

This creates value across multiple levels.

• Reduced transportation costs

• Improved delivery timelines

• Increased productivity

The system becomes not only faster—but more dependable.

New Economic Opportunities

Beyond cost reduction, the SAN introduces new forms of economic value.

Freight Access and Usage

The system generates revenue through controlled access. Freight operators utilize the network to move goods efficiently, contributing to its ongoing operation and maintenance.

Energy Integration

Through its connection with renewable energy systems, the SAN produces and distributes power. Excess energy can be redirected, creating additional revenue streams and supporting broader infrastructure.

Data Infrastructure

The tunnels themselves provide a secure environment for data transmission. Fiber-optic networks can be integrated within the system, enhancing communication infrastructure while generating economic value.

These opportunities extend the function of the SAN beyond transportation.

It becomes a multi-purpose infrastructure system.

Return on Investment

Any large-scale infrastructure project must ultimately justify itself economically.

The SAN achieves this through a combination of cost reduction and revenue generation.

Over time, the system offsets its initial investment through:

• Reduced maintenance costs

• Increased operational efficiency

• New revenue streams

The break-even point is not immediate.

But it is achievable.

And beyond that point, the system continues to generate value.

The Social Dividend

While financial returns are essential, the broader impact of the SAN extends beyond economics.

It creates a **social dividend**—a set of benefits that improve quality of life.

Air quality improves as diesel emissions decrease.

Healthcare costs are reduced as pollution declines.

Property values increase as communities become quieter and more livable.

Urban spaces are no longer dominated by freight movement.

They are reclaimed.

The Wealth of Silence

Perhaps the most profound economic impact of the SAN is one that is difficult to measure.

It is the value of what is no longer present.

The absence of noise.

The absence of congestion.

The absence of disruption.

These changes do not appear on balance sheets.

But they shape daily life.

Communities become more peaceful.

Cities become more efficient.

Time becomes more accessible.

This is the **wealth of silence**.

A form of prosperity defined not by what is built—but by what is removed.

A Generational Investment

For Dustin, Matthew, Georgia, and Jolie, the significance of this system is not measured in immediate returns.

It is measured in long-term impact.

It is the understanding that the systems we invest in today determine the conditions of tomorrow.

That infrastructure is not only a financial decision—but a societal one.

The SAN represents a commitment to building systems that last.

Systems that serve not only current needs, but future generations.

Redefining Economic Progress

For decades, progress has been associated with expansion.

More roads.

More vehicles.

More visible infrastructure.

But this model comes with increasing cost—both financial and environmental.

The SAN introduces a different perspective.

Progress is not about doing more.

It is about doing better.

It is about creating systems that are efficient, sustainable, and unobtrusive.

The Invisible Economy

In its final form, the SAN represents a new kind of economy.

An economy that operates beneath the surface.

An economy that supports the visible world without overwhelming it.
An economy built on precision, efficiency, and integration.

It is not defined by expansion, but by refinement.

The Quiet Transformation

The transformation brought by the SAN is not dramatic in appearance.

It does not rise above cities or reshape landscapes in visible ways.

It operates quietly.
But its impact is far-reaching.
The surface becomes less burdened.
Systems become more efficient.
Life becomes more balanced.

The True Value

In the end, the value of the SAN is not measured solely in financial terms.

It is measured in what it restores.

Time.
Space.
Air.
Stability.

It is the realization that progress does not have to come at a cost.

That systems can evolve.

That infrastructure can serve without burdening.

The Wealth We Carry Forward

The SAN does not eliminate the movement of goods.
It refines it.
It transforms a visible burden into an invisible foundation.

And in doing so, it creates a form of wealth that extends beyond economics.

A wealth defined by balance, sustainability, and quiet efficiency.

A wealth that future generations will inherit—not as a system they must fix, but as one that continues to serve them.

Chapter 6

Global Scalability (From Fiji to the World)

Beyond a Single Region

Every transformative system begins with a point of origin.

It is tested within a specific environment, shaped by local needs, refined through real-world conditions. But its true measure is not how well it functions in one place.

It is how far it can reach.

The Subterranean Arterial Network, proven through models such as the Golden State Sub-Arterial, is not limited by geography. It is not confined to a single region or economy.

It is a system built on principles—principles that can be applied wherever the movement of goods intersects with the limits of infrastructure.

And that intersection exists everywhere.

A Universal Challenge

Across the world, nations face the same fundamental dilemma.

Trade continues to grow. Populations expand. Demand increases. Yet the infrastructure designed to support these systems struggles to keep pace.

Highways become congested.

Urban centers become saturated.

Environmental pressures intensify.

The problem is not unique to one country.

It is global.

And while each region may experience it differently, the underlying issue remains consistent:

How do we move more—without overwhelming what we already have?

A Framework, Not a Fixed System

The SAN is not designed as a singular project to be replicated identically across every region.

It is designed as a framework.

A system defined by adaptable components, scalable structures, and universal principles. This allows each implementation to respond to local conditions while remaining aligned with a broader global network.

The SAN does not impose uniformity.

It enables compatibility.

This distinction is critical.

A rigid system cannot scale across diverse environments. A flexible framework can.

Adapting to the Landscape

One of the most significant advantages of the SAN is its ability to integrate into existing environments without competing for surface space.

Traditional infrastructure often requires extensive land use—clearing areas, displacing communities, and altering ecosystems.

The SAN operates beneath these constraints.

It adapts to the landscape rather than reshaping it.

In dense urban environments, it bypasses congestion by moving below it.

In mountainous regions, it avoids surface disruption by tunneling through stable formations.

In environmentally sensitive areas, it preserves ecosystems by minimizing surface impact.

Each implementation is unique.

But the principle remains the same:

Preserve the surface.

Optimize the system below.

The Fiji Model: Infrastructure Without Intrusion

In smaller, environmentally sensitive regions, the SAN demonstrates one of its most powerful capabilities: restraint.

Island nations such as Fiji face a unique set of challenges. Infrastructure is limited. Ecosystems are fragile. Cultural and historical landscapes are deeply intertwined with the natural environment.

In such contexts, traditional freight systems can cause disproportionate disruption.

The SAN offers an alternative.

Through the **Fiji Model**, the system is implemented as a network of smaller, localized tunnels—what can be described as **micro-arteries**.

These tunnels connect ports to inland communities, allowing goods to move efficiently without relying on heavy surface transport.

The benefits extend beyond logistics:

- Coastal roads are preserved
- Natural ecosystems remain undisturbed
- Cultural sites are protected

Development continues.

But it does so without eroding what makes these regions unique.

Scaling Up: From Islands to Continents

While the Fiji Model represents a smaller-scale implementation, the same principles apply to larger regions.

In Europe, the SAN can reduce the environmental impact of freight moving through mountainous terrain, preserving landscapes while maintaining connectivity.

In Asia, where megacities face extreme congestion, the system can bypass surface limitations entirely—delivering goods beneath dense urban environments without adding to traffic.

In North America, the GSSA serves as the foundation for a broader network—connecting regions through a system that operates independently of surface constraints.

Each application varies in scale.

But each follows the same logic.

The Network Beneath the Network

As SAN systems expand across regions, they begin to connect.

Individual networks link together, forming a larger system—a **network beneath the network.**

This subterranean layer operates independently of surface infrastructure, creating a parallel system that supports global trade without competing with it.

Goods move across regions without interruption.

Supply chains become more resilient.

Dependencies on surface conditions are reduced.

This is not simply expansion.

It is integration.

Decoupling Growth from Impact

One of the most significant implications of global SAN implementation is the ability to separate economic growth from environmental impact.

Traditionally, increased trade results in increased emissions, congestion, and infrastructure strain.

The SAN disrupts this relationship.

By relocating freight underground and powering it through renewable energy, it allows trade to expand without increasing its negative effects.

Growth continues.

Impact decreases.

This is not a compromise.

It is a redefinition.

The Role of Standardization

For a system of this scale to function globally, consistency is essential.

While the SAN is adaptable, it must also be standardized.

This ensures that systems built in different regions can connect seamlessly.

Standardization includes:

- Compatible tunnel dimensions

- Uniform transport systems

- Shared safety protocols

- Integrated data systems

This creates interoperability.

A container entering the system in one region can move through another without modification. Infrastructure built in different locations becomes part of a unified whole.

The system becomes global not through expansion alone—but through alignment.

A System Without Borders

One of the limitations of traditional infrastructure is its dependence on geography.

Road systems change across borders. Regulations differ. Compatibility becomes a challenge.

The SAN transcends these limitations.

By operating beneath the surface and adhering to universal standards, it creates a system that is not constrained by borders.

Goods move continuously.

Not because policies have been simplified—but because the infrastructure itself is unified.

The Environmental Equation

As the SAN expands, its environmental impact becomes increasingly significant.

Each implementation reduces reliance on diesel transport. Each system decreases emissions. Each network contributes to improved air quality and reduced noise pollution.

The cumulative effect is substantial.

A global SAN network would represent one of the largest shifts toward sustainable logistics in modern history.

It is not a localized improvement.

It is a planetary one.

The Human Perspective

Beyond infrastructure and economics, global scalability reflects a broader shift in how humanity approaches development.

It represents a move away from systems that impose on the environment toward systems that integrate with it.

A move from expansion to balance.

From disruption to preservation.

For future generations, this shift defines the world they inherit.

A Legacy Without Boundaries

For Dustin, Matthew, Georgia, and Jolie, the significance of this chapter lies in its scope.

It is no longer about one region or one system.

It is about a world connected through thoughtful design.

A world where infrastructure serves not only local needs, but global stability.

Where systems are built not in isolation, but as part of a larger, interconnected framework.

The Quiet Expansion

The SAN does not expand in the way traditional infrastructure does.

It does not dominate landscapes or reshape skylines.

Its growth is subtle.

It moves beneath the surface—connecting regions, supporting economies, and enabling progress without disruption.

And yet, its impact is profound.

A Future Without Friction

In its fully realized form, the SAN creates a world where movement is no longer limited by surface constraints.

Goods move across continents without interruption.

Cities grow without congestion.

Economies expand without increasing their footprint.

This is not the removal of complexity.

It is the refinement of it.

The World Above, Protected

In the end, the purpose of global scalability is not expansion for its own sake.

It is protection.

Protection of infrastructure.

Protection of the environment.

Protection of the quality of life.

The SAN does not seek to change the surface.

It seeks to preserve it.

The Silent Network

As the SAN expands across regions and continents, it becomes something remarkable:

A system that is everywhere—and yet nowhere to be seen.

It operates beneath the world, supporting it without imposing upon it.

And in that quiet presence, it fulfills its greatest promise:

To carry the weight of global progress—while leaving the world above untouched.

CHAPTER 7

The Global Standard (From California to Fiji)

From Innovation to Universality

Every transformative system must eventually answer a defining question:

Can it be shared?

Innovation, in its earliest form, is often localized—tested within a specific region, shaped by particular needs, and refined through controlled implementation. But true transformation occurs when that innovation becomes universal.

When it can move beyond its point of origin.

When it can be adopted across cultures, climates, and economies.

When it becomes not just a solution—but a standard.

The Subterranean Arterial Network has reached this threshold.

What began as a conceptual response to a regional challenge has evolved into a framework capable of global application.

The next step is not expansion alone.

It is standardization.

The Importance of a Common Language

Infrastructure, at its core, is a form of communication.

Roads connect regions.

Ports connect continents.

Rail systems connect economies.

But for these systems to function seamlessly, they must share a common language—a set of standards that ensures compatibility.

Without this shared framework, systems become fragmented.

Connections become inefficient.

Transitions become complex.

Integration becomes limited.

The history of global trade offers a powerful example of how standardization can reshape an entire industry.

The Lesson of the Shipping Container

Before the introduction of the standardized shipping container, global trade was slow, labor-intensive, and inefficient.

Goods were loaded and unloaded manually, often repackaged multiple times as they moved between ships, trains, and trucks. Each transition introduced delay, cost, and potential error.

The container changed everything.

By creating a uniform unit of transport, it allowed goods to move seamlessly across different modes of transportation. Ships, trucks, and rail systems could all accommodate the same container without modification.

This simplicity created efficiency.

Efficiency created scalability.

Scalability transformed global trade.

The SAN builds upon this principle—but extends it further.

It does not standardize the cargo.

It standardizes the system itself.

The Standardized Arterial Network

At the heart of the SAN's global vision is the concept of a **Standardized Arterial Network**.

This means that regardless of where the system is built, it follows a unified set of principles and specifications.

These include:

- Consistent tunnel dimensions and structural design

- Standardized maglev sled systems

- Unified energy integration frameworks

- Shared safety and monitoring protocols

- Interoperable data and communication systems

This level of standardization ensures that systems built in different regions can connect seamlessly.

A sled developed in one country can operate in another.

A network constructed in one region can integrate with its neighbor.

The result is not just compatibility.

It is continuity.

Modularity as the Key to Expansion

Standardization does not require uniformity in scale.

The SAN achieves flexibility through modular design.

Instead of constructing identical systems everywhere, it provides a set of standardized components that can be assembled based on local needs.

These modules include:

- Tunnel construction systems

- Transport and propulsion units

• Sensor and monitoring networks

• Energy integration components

• Hub and distribution frameworks

Each module follows a shared design, ensuring compatibility while allowing for variation in size and capacity.

This approach allows regions to implement the SAN in a way that reflects their specific conditions—while remaining connected to the global network.

From California to Fiji: A Unified System

The transition from large-scale systems such as the GSSA to smaller implementations like the Fiji Model demonstrates the flexibility of this approach.

In California, the SAN operates as a major logistical backbone—connecting ports, hubs, and regional networks across vast distances.

In Fiji, the system takes a different form.

It becomes a network of smaller, localized tunnels—designed to preserve the environment while improving connectivity.

Despite these differences, both systems share the same underlying framework.

They are not separate solutions.

They are variations of the same system.

This is the power of standardization.

Crossing Borders Without Friction

One of the most significant limitations of traditional infrastructure is its dependence on national boundaries.

Road systems change from one country to another. Regulations vary. Compatibility becomes a challenge.

These differences create friction.

The SAN eliminates this friction.

By operating beneath the surface and adhering to universal standards, it creates a system that transcends borders.

Goods move continuously.

There are no interruptions caused by incompatible systems. No delays caused by transitions between infrastructure types.

The network functions as a unified whole.

The Role of Data Integration

In addition to physical standardization, the SAN relies on a unified data framework.

Every component of the system—sleds, sensors, hubs, and energy systems—communicates through a shared network.

This creates a consistent flow of information across regions.

Data is exchanged in real time.

System performance is monitored continuously.

Adjustments are made dynamically.

This integration ensures that the system operates not as a collection of parts, but as a coordinated whole.

It becomes a network that not only moves goods—but understands how it moves them.

Sustainability as a Global Standard

Standardization is not only about compatibility.

It is also about consistency in impact.

The SAN establishes a new global benchmark for sustainable infrastructure.

Each implementation is designed to:

• Reduce reliance on fossil fuels

• Minimize environmental disruption

• Improve energy efficiency

• Lower emissions

As the system expands, these benefits accumulate.

The result is a global infrastructure network that supports economic growth while reducing environmental impact.

A Shared Responsibility

For the SAN to become a global standard, it requires more than engineering.

It requires collaboration.

Governments, industries, and communities must align around a shared vision of infrastructure—one that prioritizes efficiency, sustainability, and long-term value.

This is not a challenge.

It is an opportunity.

An opportunity to build systems that serve not only individual regions, but the world as a whole.

The Emergence of Silent Trade

As the SAN becomes standardized and interconnected, it gives rise to a new form of global exchange.

Trade that is continuous, efficient, and nearly invisible.

Goods move across regions without disruption.

Supply chains operate without delay.

Communities remain unaffected by the movement that sustains them.

This is the emergence of **Silent Trade**.

A system where the movement of goods no longer defines the spaces in which people live.

A Legacy Without Boundaries

For Dustin, Matthew, Georgia, and Jolie, the significance of this chapter lies in its reach.

It represents a world where infrastructure is no longer confined by geography.

A world where systems are designed with compatibility in mind from the very beginning.

A world where progress is shared.

The Infrastructure of Unity

The SAN, once standardized, becomes more than a network.

It becomes a foundation for global connection.

A system that supports not only trade, but communication, innovation, and collaboration.

It is infrastructure not as a limitation—but as an enabler.

The Quiet Integration

The expansion of the SAN does not reshape skylines or redefine landscapes.

It integrates quietly.

Beneath cities.

Beneath borders.

Beneath the systems that already exist.

Its presence is subtle.

Its impact is significant.

A World Connected Differently

In its final form, the SAN creates a new model of global connectivity.

One that is not defined by visible infrastructure, but by invisible precision.

Goods move seamlessly across continents.

Systems operate without interruption.

Communities thrive without disruption.

This is not the future of infrastructure as we have known it.

It is the evolution of it.

The Silent Connection

The SAN does not seek recognition.

It seeks function.

It operates beneath the surface, connecting regions, supporting economies, and enabling progress without drawing attention to itself.

And in that quiet operation, it achieves something extraordinary:

It brings the world closer together—without changing the world above it.

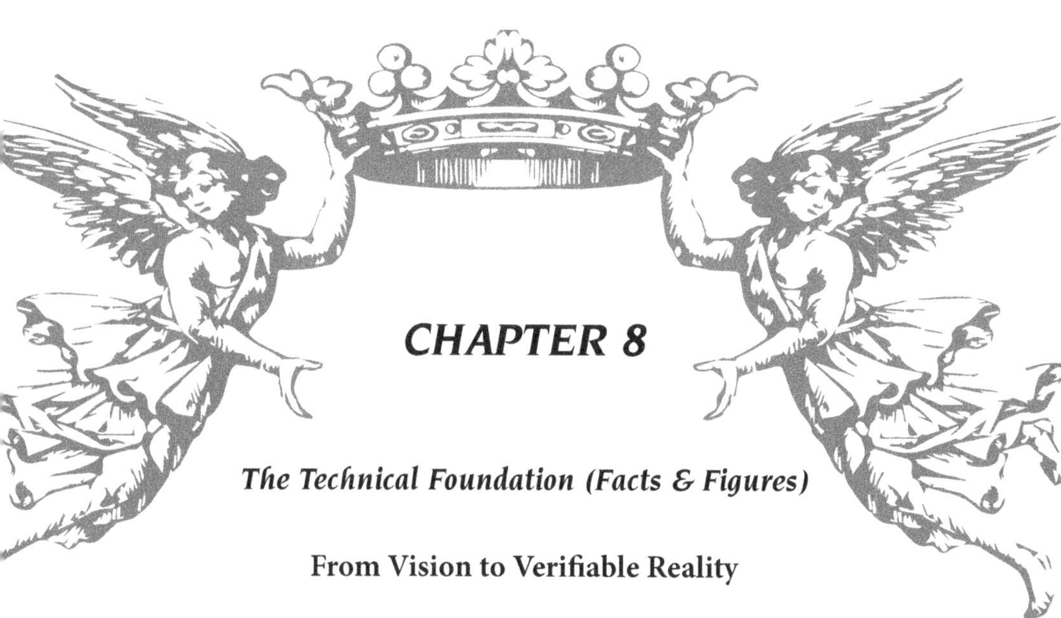

CHAPTER 8

The Technical Foundation (Facts & Figures)

From Vision to Verifiable Reality

Every transformative idea must eventually confront a fundamental requirement:

It must be measurable.

Vision inspires. Philosophy guides. But it is precision—numbers, materials, and specifications—that determines whether a system can be built, trusted, and sustained.

The Subterranean Arterial Network is not defined by imagination alone. It is grounded in engineering realities—systems designed not only to function, but to perform consistently under the demands of scale, time, and environment.

This chapter presents the technical foundation of the SAN—the components, metrics, and systems that translate vision into implementation.

Because in the end, progress is not proven by what is proposed.

It is proven by what can be built.

The Core Unit: The Maglev Sled

At the center of the SAN's operation is its most fundamental element: the **maglev sled**.

These sleds are the carriers of the system—the units responsible for transporting freight continuously through the subterranean network. Each sled is designed to balance strength, efficiency, and durability.

Unlike traditional freight vehicles, which rely on mechanical contact and experience constant wear, the sled operates within a frictionless environment.

It does not roll.

It levitates.

Using **Electrodynamic Suspension (EDS)**, the sled is lifted above the track by magnetic forces. This eliminates physical contact, significantly reducing mechanical degradation.

Propulsion is achieved through **Linear Induction Motors (LIM)**, which generate electromagnetic fields that move the sled forward.

This combination creates a system that is:

- Smooth in operation

- Efficient in energy use

- Minimal in maintenance requirements

The sled is not constrained by the limitations of traditional transport.

It operates within a different paradigm.

Capacity and Performance

The sled is designed to handle the demands of modern freight.

It accommodates standardized containers, ensuring compatibility with existing global logistics systems. Its structure supports high payload capacities while maintaining stability at speed.

Performance is defined not only by how much it can carry, but by how consistently it can move.

Unlike surface transport, where speed fluctuates, the sled maintains a steady velocity within the controlled environment of the SAN.

This consistency is critical.

It transforms logistics from a variable system into a predictable one.

Energy Efficiency and Recovery

Efficiency within the SAN is achieved not only through movement, but through energy management.

One of the key features of the maglev system is **regenerative braking**.

When a sled decelerates, the energy generated is captured and redistributed within the network. This reduces overall energy consumption and enhances system sustainability.

Energy is not wasted.

It is reused.

This approach aligns with the broader philosophy of the SAN—maximizing efficiency at every level.

Tunnel Engineering: The Structural Environment

If the sled represents motion, the tunnel represents stability.

The tunnel is not simply a passageway.

It is a controlled environment designed to support continuous operation over extended periods.

Construction materials are selected for durability, resilience, and longevity.

• **High-strength composite linings** provide structural integrity

• **Nano-concrete materials** enhance resistance to wear and corrosion

• **Polymer-based seismic joints** allow flexibility under stress

These materials create a structure capable of enduring both environmental forces and operational demands.

Unlike surface infrastructure, which is exposed to weather, temperature fluctuations, and constant load variation, the SAN operates within a stable environment.

This stability extends its lifespan significantly.

Environmental Control

The tunnel environment is carefully managed to optimize performance.

Temperature, pressure, and air density are regulated to ensure efficient operation.

The introduction of a **low-pressure environment** reduces air resistance, allowing sleds to move at higher speeds with less energy.

This controlled atmosphere is maintained through integrated systems that monitor and adjust conditions in real time.

The result is an environment that is not only efficient, but predictable.

The Sensor Network: Awareness and Control

Embedded throughout the SAN is a network of sensors that provide continuous awareness of system conditions.

These sensors operate across multiple domains:

- **Acoustic sensors** detect structural changes

- **Thermal sensors** monitor heat variations

- **Pressure sensors** ensure environmental stability

- **Vibration sensors** identify mechanical and seismic activity

Together, they create a system that is constantly observing itself.

This awareness is essential.

It allows the system to respond to changes before they become problems.

Data Integration and Communication

The SAN is not only a physical system—it is a digital one.

Data flows alongside cargo, creating a continuous stream of information that supports system operation.

Each sled communicates with the network, providing real-time updates on position, speed, and performance.

This information is processed continuously, allowing the system to adjust dynamically.

Movement is coordinated.

Performance is optimized.

Efficiency is maintained.

The system does not operate blindly.

It operates intelligently.

Predictive Maintenance: Anticipating Failure

One of the defining features of the SAN is its ability to anticipate issues before they occur.

Through continuous data analysis, the system identifies patterns that indicate potential future problems.

A slight increase in vibration.

A gradual rise in temperature.

A minor shift in alignment.

These signals trigger maintenance actions before failure occurs.

Autonomous repair units are deployed to address issues proactively, ensuring that the system remains operational without interruption.

Maintenance becomes a continuous process—not a reaction to breakdown.

System Integration: A Unified Whole

The strength of the SAN lies in its integration.

Each component—sleds, tunnels, sensors, and energy systems—does not function independently.

They operate as a unified system.

Energy powers movement.

Sensors monitor conditions.

Data informs decisions.

This integration creates efficiency that cannot be achieved through isolated systems.

It ensures that every element contributes to the performance of the whole.

Performance Metrics

The effectiveness of the SAN can be measured through key performance indicators.

These include:

- Reduced energy consumption per unit of freight
- Increased consistency in delivery times
- Extended infrastructure lifespan
- Lower operational and maintenance costs

These metrics demonstrate that the system is not only innovative—but practical.

It delivers measurable improvements.

Durability Over Time

The SAN is designed with a long-term perspective.

Each component is evaluated for its performance over decades—not just years.

Tunnel structures are built to endure environmental stress.

Maglev systems are designed for sustained operation with minimal wear.

Sensor networks are engineered for continuous monitoring.

The system is not intended to be replaced.

It is intended to be maintained.

The Science of Precision

For Dustin, Matthew, Georgia, and Jolie, this chapter represents the underlying truth of the system:

That large-scale change is built on precise details.

It is not the size of the system that determines its success.

It is the accuracy of its design.

Sensors that detect microscopic changes.

Materials engineered for long-term resilience.

Systems calibrated for continuous performance.

This precision ensures reliability.

The Invisible Performance

The greatest achievement of the SAN's technical foundation is not its complexity.

It is its invisibility.

The system operates continuously, yet remains unseen. It performs at a high level, yet does not intrude upon daily life.

Its success is measured not by its presence—but by its absence.

The absence of failure.

The absence of disruption.

The absence of inefficiency.

A Foundation That Endures

The SAN represents a shift in how infrastructure is built.

It is no longer temporary.

It is no longer reactive.

It is designed to endure.

A system that supports not only current needs, but future generations.

A system that exists beneath the surface—quietly, reliably, and continuously.

The Final Measure

In the end, the value of the SAN's technical foundation lies in its certainty.

The certainty that it will function as intended.

The certainty that it will adapt to challenges.

The certainty that it will endure over time.

Because true progress is not defined by what is imagined.

It is defined by what continues to work—quietly, precisely, and without interruption.

Conclusion

The Silent Renaissance

The End of the Visible Burden

For generations, humanity has lived with a compromise.

We accepted that progress would be loud.

That growth would be disruptive.

That the movement of goods—the very engine of civilization—would come at the cost of our environment, our infrastructure, and our daily lives.

We built systems that worked.

But we also built systems that intruded.

Highways became congested.

Cities became louder.

Air became heavier.

And over time, these conditions were no longer seen as problems to be solved—but as realities to be endured.

This was the "Worthy Nuisance."

A necessary burden.

An accepted trade-off.

Until now.

Redefining Progress

The Subterranean Arterial Network challenges a belief that has shaped modern infrastructure:

That progress must always be visible.

To move the world, we must see the machinery that moves it. That efficiency requires expansion across the surface.

But what if progress could take a different form?

What if the systems that sustain our lives could operate without overwhelming them?

What if infrastructure could exist—not as a presence, but as a foundation?

The SAN answers these questions not with theory, but with design.

It does not remove the need for movement.

It removes the disruption caused by it.

A World Reclaimed

When the weight of global logistics is moved beneath the surface, something remarkable happens above it.

The surface is returned.

Roads become smoother and last longer.

Traffic becomes lighter and more predictable.

Cities grow quieter.

Air becomes cleaner.

Spaces once dominated by freight movement begin to transform.

Communities reclaim their surroundings.

Urban areas regain balance.

Nature is given room to recover.

This is not a reduction in progress.

It is an enhancement of it.

The Economy of Balance

The SAN does more than improve infrastructure—it redefines the relationship between growth and impact.

For the first time, economic expansion is no longer tied to increased strain on physical systems.

Trade can grow.

Movement can increase.

Demand can expand.

Without adding congestion.

Without accelerating deterioration.

Without increasing environmental harm.

This is the emergence of a new kind of economy—one that operates efficiently, sustainably, and quietly.

An economy built not on visible expansion, but on invisible precision.

The Invisible System

Perhaps the most profound aspect of the SAN is not what it builds, but what it removes.

It removes noise.

It removes congestion.

It removes pollution.

And in doing so, it creates something that is rarely associated with infrastructure:

Silence.

Not the absence of activity, but the absence of disruption.

A system that continues to function—constantly, reliably, and efficiently—without demanding attention.

This is the Silent Renaissance.

A return not to the past, but to balance.

A Legacy for the Future

For Dustin, Matthew, Georgia, and Jolie, this is more than an infrastructure proposal.

It is a promise.

A promise that the systems built today will not burden their tomorrow. A promise that progress will no longer come at the cost of their environment. A promise that the world they inherit will be cleaner, quieter, and more sustainable.

It is the understanding that innovation is not only about advancement—but about responsibility.

The Shift Beneath Our Feet

The transformation proposed by the SAN is not dramatic in appearance.

There are no towering structures.

No visible expansion across landscapes.

Instead, the change happens quietly.

Beneath roads.

Beneath cities.

Beneath the very ground we walk on.

It is a shift not in what we see, but in how systems operate.

And in that shift lies its power.

The Final Perspective

For over a century, we believed that to move the world, we had to see and hear the movement.

We believed that progress would always leave a mark.

But the greatest advancement of the next era may be defined by the opposite:

A system so efficient, so precise, and so integrated that it disappears from view.

A system that supports everything—while demanding nothing.

The Silent Renaissance

This is not the end of industry.

It is its evolution.

A transition from visible burden to invisible foundation.

From disruption to harmony.

From compromise to design.

The Subterranean Arterial Network is not simply a solution to a problem.

It is a redefinition of how progress itself is achieved.

And as it expands—quietly, steadily, and beneath the surface—it carries with it the promise of a future where the world above is no longer shaped by the burdens below.

A Final Thought

We did not remove the work.

We refined where it happens.

We did not stop the movement of the world.

We gave it a better path.

APPENDICES

Technical Blueprint & Specifications

Appendix A: Structural Dimensions

This section outlines the foundational physical specifications of the Subterranean Arterial Network (SAN), defining the structural parameters required for durability, safety, and long-term operation.

• Primary Tunnel Diameter:

Designed to accommodate maglev freight systems and maintenance access, ensuring sufficient clearance for high-capacity transport.

• Segment Thickness:

Constructed using high-strength composite and nano-concrete materials, providing resistance to pressure, corrosion, and long-term wear.

• Minimum Depth Requirements:

Urban Areas: Engineered to operate safely beneath dense infrastructure

Mountainous or Variable Terrain: Increased depth to ensure stability and environmental protection

• Track Configuration:

Standardized maglev pathway optimized for high-speed, low-resistance freight movement

Appendix B: Power & Propulsion Systems

This section details the energy and motion systems that enable efficient and sustainable operation of the SAN.

• Propulsion System:

Linear Induction Motors (LIM) providing smooth, high-speed acceleration without mechanical contact

• **Energy Source:**

Direct integration with renewable energy systems (SHIELD Project), ensuring sustainable and continuous operation

• **Electrical System:**

Direct current (DC) transmission minimizing energy loss and maximizing efficiency

• **Maglev Suspension:**

Electrodynamic suspension (EDS) allowing frictionless movement and reduced maintenance

• **Energy Recovery:**

Regenerative braking systems capturing and redistributing energy within the network

Appendix C: Safety Systems (SISS)

SHIELD Integrated Shutdown System

This section outlines the safety architecture designed to ensure continuous operation and rapid response to anomalies.

• **Acoustic Monitoring:**

Fiber-optic sensing systems detecting structural irregularities and mechanical disturbances

• **Thermal Detection:**

Continuous monitoring of heat levels to identify potential failures or fire risks

• **Pressure Monitoring:**

Maintaining controlled tunnel environments and identifying breaches

• Response Time:

Near-instantaneous detection and isolation of affected segments

• Isolation Mechanism:

Segment-specific shutdown using high-strength containment systems

• Automated Response:

Deployment of autonomous repair units for immediate issue resolution

Appendix D: Environmental & Economic Metrics

This section presents key performance indicators demonstrating the SAN's impact on sustainability and economic efficiency.

Environmental Impact

• Significant reduction in carbon emissions through elimination of diesel-based freight transport

•Improved air quality in urban and industrial regions

• Reduced noise pollution across transportation corridors

Infrastructure Impact

• Extended lifespan of roads and bridges due to reduced heavy-load stress

• Decreased frequency of maintenance and repair cycles

• Improved reliability of transportation systems

Economic Impact

• Reduced operational costs for freight movement

• Increased efficiency and predictability in supply chains

• New revenue streams through integrated energy and data systems

Long-Term Value

The SAN is designed as a generational infrastructure investment, delivering:

• Sustainable economic growth

• Environmental preservation

• Improved quality of life

Appendix E: Master System Summary

For clarity and reference, the core components of the SAN framework are summarized below:

• **Primary System:** Subterranean Arterial Network (SAN)

• **Regional Model:** Golden State Sub-Arterial (GSSA)

• **Safety Framework:** SHIELD Integrated Shutdown System (SISS)

• **Energy Source:** Renewable SHIELD Project systems

• **Operational Focus:** Fully automated, freight-only underground transport